Earth's Changing Surface
Surface
Rocks and Minerals

by Kate Boehm Jerome

Table of Contents

Develop Language 2

CHAPTER 1   Properties of Rocks and Minerals . . 4

Your Turn: Communicate 9

CHAPTER 2   How Rocks Form 10

Your Turn: Summarize 15

CHAPTER 3   How Rocks Break Down and Move 16

Your Turn: Interpret Data 19

Career Explorations 20

Use Language to Ask Questions 21

Science Around You 22

Key Words 23

Index 24

Millmark
EDUCATION

The Crater of Diamonds State Park is in Murfreesboro, Arkansas. People who visit can dig for diamonds and other **minerals**.

Discuss the photos of minerals with questions like these:

What do the diamonds look like?

The diamonds look _____.

What color is the calcite?

The calcite is _____.

How does the banded agate compare to the other minerals?

The banded agate is _____.

How are the minerals different?

mineral – a natural solid element or compound with a definite structure

calcite

peridot

banded agate

diamonds

MISSOURI

0   50   100 Miles

0   50   100 Kilometers

ARKANSAS

TENN.

Little Rock ★

OKLAHOMA

Murfreesboro ●

MISSISSIPPI

Crater of Diamonds
State Park

UNITED
STATES

TEXAS

LOUISIANA

garnet

CHAPTER 1

Properties of Rocks and Minerals

Have you ever wondered what rocks are made of? Rocks are made of one or more minerals. A mineral is a solid. Some minerals are elements and are made of only one kind of **atom**. Other minerals are compounds and are made of at least two different elements joined together.

Each mineral has a definite **structure** and is formed naturally. A mineral is not alive and it does not form from anything that was once alive. Gold is a mineral. But coal is not a mineral because it forms from decaying plant and animal material.

atom – the smallest whole unit of matter
structure – the orderly arrangement of atoms

A rock can be made of just one mineral. But most rocks are made of a **mixture** of minerals. Different minerals come together to form a specific type of rock. Look at the granite rock on this page. You can see that there are different minerals in it.

▼ Granite is a mixture of minerals.

mixture – something made of two or more different things

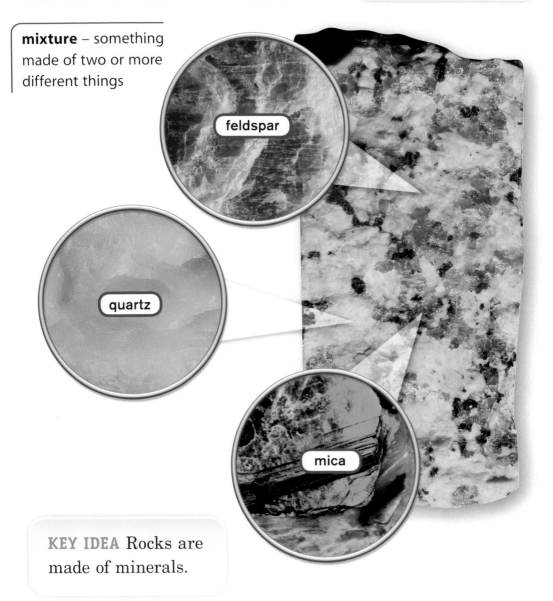

feldspar

quartz

mica

KEY IDEA Rocks are made of minerals.

Identifying Minerals

Look at the rocks in these pictures. Each rock is made of the mineral quartz. How would you describe these types of quartz? You might say they are pink, purple, orange, and brown. When you describe the colors of the quartz, you are telling about its **properties**.

Properties help us identify, or name, minerals. For example, color is a property of minerals. The rose quartz mineral in the picture has a pink color.

But some minerals, such as quartz, come in many colors. Also, different minerals can sometimes have the same color. For example, the minerals gold and pyrite have almost the same color. For these reasons, sometimes other properties must be used to identify minerals.

properties – qualities that can be observed or measured

rose quartz

amethyst

citrine

smoky quartz

▶ **These types of quartz all have different colors.**

Some minerals are identified by a property called **streak**. A streak is the powdery mark that some minerals leave behind when they are rubbed on special tile. The color of the streak may be different from the color of the mineral. For example, if you rub gold and pyrite across white tile, each leaves a different streak of color. The streaks help tell these two minerals apart.

pyrite

gold

Another property called **luster** tells how a mineral reflects light. **Hardness** is a property of minerals that measures how easily a mineral can be scratched.

▲ **Gold and pyrite leave different streaks.**

streak – the powdery mark left when some minerals are rubbed across a surface

luster – a property that describes how a mineral reflects light

hardness – a property of minerals that measures how easily a mineral can be scratched

By The Way...

Pyrite is often called "fool's gold" because so many people are fooled into thinking it is real gold.

diamond

graphite

◀ **Graphite has a shiny luster.**

▲ **A diamond is the hardest mineral.**

Identifying Rocks

You know that rocks, such as granite, can be identified by the minerals they contain. But rocks can be identified in other ways, too.

Sometimes rocks can be identified by how the pieces of mineral in a rock fit together. Many times, the minerals in rocks fit together to make the rock look like one solid piece. Other times, a rock looks like it is made of chunks or pieces that are stuck together.

Texture also helps identify some rocks. Texture is based on the size and shape of the material that makes up the rock. If a rock looks and feels bumpy, its texture is rough. If it looks and feels smooth, its texture is glassy.

texture – a property that is based on the size and shape of the material making up the rock

Minerals

Granite has minerals that are easily seen.

Fit

Conglomerate rocks look chunky.

Texture

Obsidian is a rock with a glassy texture.

KEY IDEA Properties help identify minerals and rocks.

YOUR TURN

COMMUNICATE

Look at the pictures of pyrite and gold on page 7. With a friend, take turns answering these questions.

1. Why can't you use the property of color to tell these minerals apart?

The color of these two minerals _____.

2. How can you tell pyrite from gold?

A _____ test can help tell these minerals apart.

3. Why is pyrite called fool's gold?

Pyrite is called fool's gold because _____.

MAKE CONNECTIONS

Diamonds are sometimes used in the tips of drilling machines that dig through many layers of rock. Tell why you think a diamond is good for this job.

USE THE LANGUAGE OF SCIENCE

What are some of the properties that can be used to identify minerals?

Color, streak, luster, and hardness are properties that can be used to identify minerals.

CHAPTER 2

How Rocks Form

sedimentary

metamorphic

igneous

You have seen how properties help identify rocks. But where do rocks get these properties? Rocks get their properties from how they are formed. Rocks are classified into three main groups according to how they form.

The three groups of rock are **igneous rock**, **sedimentary rock**, and **metamorphic rock**. All rocks can be put into one of these three groups.

igneous rock – rock formed when hot, melted rock cools

sedimentary rock – rock formed when tiny pieces of rock and other particles get squeezed together

metamorphic rock – rock formed when extreme heat and pressure change one type of rock into another

KEY IDEA Rocks are classified according to how they form.

Igneous rock forms when melted rock, or **magma**, begins to cool. Hot magma rises from within Earth. As it makes its way toward the surface, it cools and hardens. Igneous rock can form underground in this way.

Sometimes magma reaches the surface of Earth through a volcano. The melted rock that comes out of a volcano is called **lava**. As lava cools and hardens, igneous rock is formed above ground.

magma – hot, melted rock under Earth's surface

lava – hot, melted rock that reaches Earth's surface

Explore Language

Igneous is from a Latin word, *ignis*, which means "fire".

▼ **When lava cools, igneous rock is formed.**

Sediments Pile Up

Sedimentary rock forms in a different way. Over thousands and thousands of years, little bits of rock are broken down and carried away. These little bits of rock and other sediments begin to pile up in layers. Over a long period of time, the weight of the top layers puts pressure on the bottom layers. The bottom layers begin to stick together and then harden into sedimentary rock.

Chalk is a sedimentary rock formed from tiny parts of living things that once lived in the oceans. Over millions of years, the tiny parts piled up on the ocean floor and formed chalk.

sediments – tiny pieces of rock and other particles that are carried from one place to another

▼ Fossils, or signs of life in the past, are often found in sedimentary rock.

▼ The chalk cliffs in Dover, England are sedimentary rock.

SHARE IDEAS **Tell** why you think sedimentary rock usually holds the best fossils.

Heat and Pressure

Sometimes extreme heat and pressure can change rocks. Chemical processes within Earth can also cause change. When one type of rock changes into another type of rock, metamorphic rock forms. Metamorphic rock can form from igneous, sedimentary, or even other metamorphic rocks.

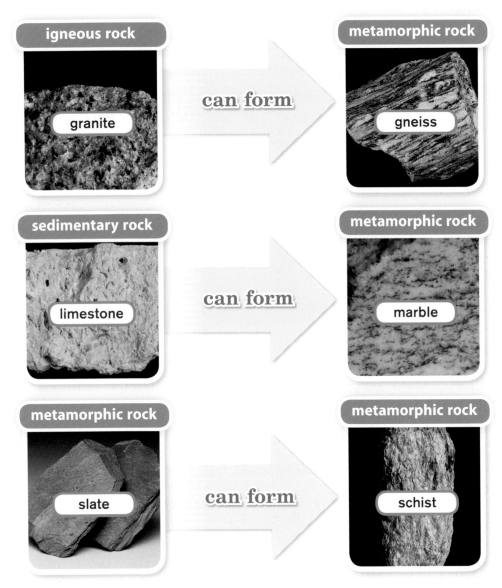

igneous rock — granite **can form** metamorphic rock — gneiss

sedimentary rock — limestone **can form** metamorphic rock — marble

metamorphic rock — slate **can form** metamorphic rock — schist

The Rock Cycle

One of the most interesting things about rocks is that they are constantly changing. On the surface of Earth, rocks are constantly breaking down and being moved through **weathering** and **erosion**. Deep within Earth, rocks are constantly melting and going through other changes caused by heat, pressure, and chemical processes.

This never-ending cycle of change is called the **rock cycle**. Although it happens over a very long period of time, the rock cycle means that rocks never stay the same.

weathering – how rocks break down and change

erosion – the movement of rocks and other particles from one place to another

rock cycle – the process that happens over a long period of time in which one type of rock changes into another type of rock

KEY IDEA Rocks constantly change from one type to another in the rock cycle.

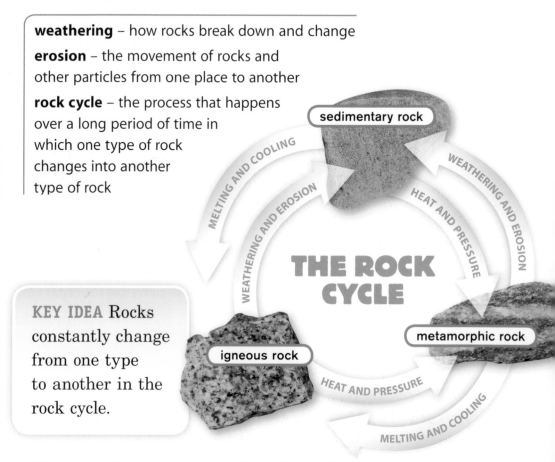

SUMMARIZE

Think about how igneous, sedimentary, and metamorphic rocks form. Summarize each process by finishing the sentences below.

1. Igneous rock forms when _____ .

2. Sedimentary rock forms when _____ .

3. Metamorphic rock forms when _____ .

MAKE CONNECTIONS

Igneous rock that is formed from a volcano is pretty easy to find. Why do you think this is so?

 STRATEGY FOCUS

Synthesize

Reread the ideas on page 14 and look at the rocks. Add what you already know about rocks. Make one statement that includes most of the information.

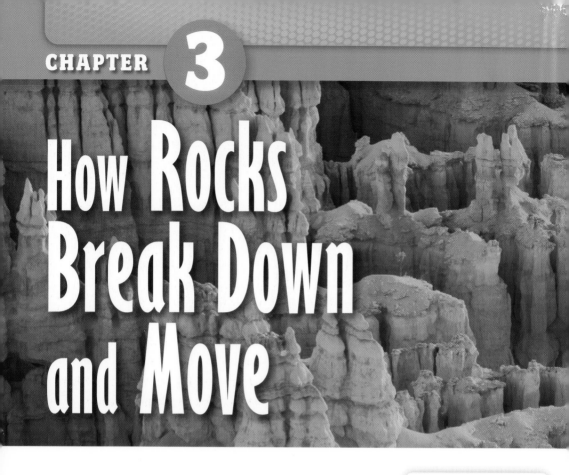

CHAPTER 3

How Rocks Break Down and Move

You cannot see how rocks are changing deep within Earth. But you can see some of the changes that happen on Earth's surface.

Rocks are constantly being broken down through the process of weathering. But there are different ways weathering can occur.

Mechanical weathering changes the size or shape of a rock. For example, when a rock is worn away by the force of water, mechanical weathering is taking place.

▲ **Different types of weathering change the rocks in Bryce Canyon, Utah.**

mechanical weathering – the breaking down of a rock into smaller pieces without changing its composition

▲ **Acid rain is causing chemical weathering of this statue.**

Chemical weathering changes the chemical composition of a rock. Water is usually involved in chemical weathering. For example, when air becomes polluted, chemicals can mix with water in the air. Acid rain can form. When acid rain falls on a marble statue, the acid rain can dissolve or change certain minerals in the marble. Over time, the statue looks different because of the chemical weathering.

Chemical and mechanical weathering often happen at the same time. The rocks you see in Bryce Canyon were formed from both types of weathering.

chemical weathering – the breaking down of a rock because of changes in its composition

Natural Forces

Natural forces cause weathering and erosion on Earth's surface. Wind, water, gravity, and glaciers can break down or move even large rocks. It is all part of the never-ending rock cycle that is always changing the surface of Earth.

▲ Wind carries particles.

▲ Water can change shorelines.

▲ Glaciers affect the land under them.

▲ Gravity can move rocks.

KEY IDEAS Weathering and erosion change Earth's surface. These changes are caused by natural forces such as wind, water, gravity, and glaciers.

YOUR TURN

INTERPRET DATA

The bar graph below shows how the beach in front of one hotel has changed over the years. Use the graph to answer the questions below.

1. In what year was there the most amount of beach?

2. In what year was there the least amount of beach?

3. How much less beach was there in 2007 than in 1977?

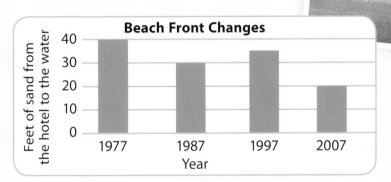

Beach Front Changes

MAKE CONNECTIONS

Tell how the erosion of sand can change a beach.

EXPAND VOCABULARY

You have read about the rock cycle. The word **cycle** comes from the Greek word *kyklos*, which means "wheel" or "circle." Look at these words. Find out what each word means, and tell how they relate to circles.

bicycle　　　　　　**recycle**　　　　　　**cyclical**

Gemologist

Look at the beautiful jewelry in the picture. It looks very valuable. But only a gemologist can tell for sure.

Gemologists study valuable minerals called gemstones. They are trained to recognize the best gemstones. They often work in jewelry stores to help design and evaluate jewelry.

Gemologists usually get formal training while earning a gemology diploma. However, work experience is also very important to this career. It usually takes years of on-the-job training for a person to become a skilled gemologist.

Do you think you would like to become a gemologist? Tell why or why not.

Use *Why, How, and What*

You can ask questions to learn more about a topic. You can begin your questions with words like **why**, **how**, and **what**.

EXAMPLES

Why does the rock cycle never end?

How can you tell the difference between pyrite and gold?

What forces can cause erosion?

With a friend, ask questions about the minerals on pages 6-7. Can you answer any of your questions?

Write Questions and Answers

This book contains many photos of rocks and minerals. Choose two rocks or minerals pictured in the book. Then ask questions about each one. Research to find the answers. Include questions and answers about the properties of the rocks or minerals.

Words You Can Use

Why is …
How does …
What is …

You might be surprised at how many minerals are around you right now. Do you have change in your pocket? The coins are made from minerals. Did you use a pencil? Another mineral is in use!

Look at the chart and answer the questions.

- What mineral might bring electricity to your computer and TV?
- What mineral is in the food you eat everyday?

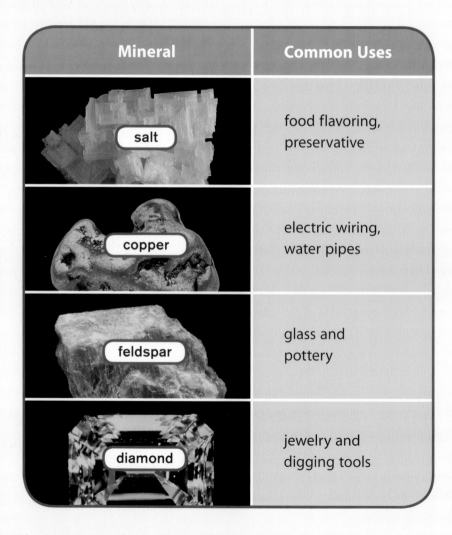

Mineral	Common Uses
salt	food flavoring, preservative
copper	electric wiring, water pipes
feldspar	glass and pottery
diamond	jewelry and digging tools

Key Words

chemical weathering the breaking down of a rock because of changes in its composition
Acid rain can cause **chemical weathering**.

erosion the movement of rocks and other particles from one place to another
Gravity can cause **erosion** by pulling loose rocks down a cliff.

igneous rock (igneous rocks) rock formed when hot, melted rock cools
Igneous rock forms when magma or lava cools.

lava hot, melted rock that reaches Earth's surface
When a volcano erupts, **lava** can shoot high in the sky.

magma hot, melted rock under Earth's surface
Magma cools as it moves toward Earth's surface.

mechanical weathering the breaking down of a rock into smaller pieces without changing its composition
Mechanical weathering can change a rock's shape.

metamorphic rock (metamorphic rocks) rock formed when extreme heat and pressure change one type of rock into another
Metamorphic rock can form deep underground.

mineral (minerals) a natural, solid element or compound with a definite structure
Diamond is the hardest **mineral**.

property (properties) a quality that can be observed or measured
Rocks and minerals can be identified by their **properties**.

rock cycle the process that happens over a long period of time in which one type of rock changes into another type of rock
The **rock cycle** is a never-ending process of change.

sedimentary rock (sedimentary rocks) rock formed when tiny pieces of rock and other particles get squeezed together
Fossils are sometimes found in **sedimentary** rock.

structure the orderly arrangement of atoms
Minerals have a definite **structure**.

weathering how rocks break down and change
Water, wind, and glaciers can cause **weathering**.

Index

atom 4

chemical weathering 17

erosion 14, 18–19

hardness 7

igneous rock 13–15

lava 11

luster 7

magma 11

mechanical weathering 16–17

metamorphic rock 8, 10, 13–15

minerals 2–9, 20, 22

properties 6–10

rock cycle 14, 18

sedimentary rock 10, 12–15

sediments 12

streak 7

structure 4

texture 8

weathering 14, 16–18

MILLMARK EDUCATION CORPORATION
Ericka Markman, President and CEO; Karen Peratt, VP, Editorial Director; Rachel L. Moir, Director, Operations and Production; Mary Ann Mortellaro, Science Editor; Amy Sarver, Series Editor; Betsy Carpenter, Editor; Guadalupe Lopez, Writer; Kris Hanneman and Pictures Unlimited, Photo Research

PROGRAM AUTHORS
Mary Hawley; Program Author, Instructional Design
Kate Boehm Jerome; Program Author, Science

BOOK DESIGN Steve Curtis Design

CONTENT REVIEWER
Tom Nolan, Operations Engineer, NASA Jet Propulsion Laboratory, Pasadena, CA

PROGRAM ADVISORS
Scott K. Baker, PhD, Pacific Institutes for Research, Eugene, OR
Carla C. Johnson, EdD, University of Toledo, Toledo, OH
Donna Ogle, EdD, National-Louis University, Chicago, IL
Betty Ansin Smallwood, PhD, Center for Applied Linguistics, Washington, DC
Gail Thompson, PhD, Claremont Graduate University, Claremont, CA
Emma Violand-Sánchez, EdD, Arlington Public Schools, Arlington, VA (retired)

PHOTO CREDITS Cover © Richard Price/Getty Images; 1 © Frank Krahmer/Getty Images; 2-3 © Eric Nathan/Alamy; 2 © GC Minerals/Alamy; 3a, 5b, 6b, 6c, 22c, © Mark A. Schneider/Photo Researchers, Inc.; 3c map by Mapping Specialists; 3d © United States Geological Survey; 3b, 5a, 5c, 7a, 7b, 7d, 7e, 8a, 13a, 13e, 13f © Dr. Richard Busch; 5d © Charles D. Winters/Photo Researchers, Inc.; 6a © Arnold Fisher/Photo Researchers, Inc.; 6d © TH Foto-Werbung/Photo Researchers, Inc.; 7c and 21 © Steffen Foerster Photography/Shutterstock; 7f © Eleonora Kolomiyets/Shutterstock;

8b © Glenn Oliver/Visuals Unlimited; 8c © Roberto Benzi/age fotostock; 8d © Tai Power Seeff/Getty Images; 9a and 9b Lloyd Wolf for Millmark Education; 10b © Joyce Photographics/Photo Researchers, Inc.; 10c © sciencephotos/Alamy; 11 © G. Brad Lewis/age fotostock; 12a © Phil Degginger/Carnegie Museum/Alamy; 12b © Peter Phipp/age fotostock; 13b © Dirk Wiersma/Photo Researchers, Inc.; 13c © E.R. Degginger/Photo Researchers, Inc.; 13d © Michael Szoenyi/Photo Researchers,Inc.; 4, 10a, 14a, 14b, 14c © Wally Eberhart/Visuals Unlimited; 15 © Andoni Canela/age fotostock; 16-17 © Jamie and Judy Wild/Danita Delimont; 17 © Cordelia Molloy/Photo Researchers, Inc.; 18a © Chris Anderson/Aurora/Getty Images; 18b © R. Matina/age fotostock; 18c © Marli Miller/Visuals Unlimited; 18d © Jean-Pierre Clatot/AFP/Getty Images; 19 © Avner Richard/Shutterstock; 20a © Mike Goldwater/Alamy; 20b © Photodisc/Punchstock; 22a © Kevin Schafer/Peter Arnold, Inc.; 22b © Kaj R. Svensson/Photo Researchers, Inc.; 22d © V. Fleming/Photo Researchers, Inc.; 24 © Justin Kim/Shutterstock

Copyright © 2008 Millmark Education Corporation

Published by Millmark Education Corporation
7272 Wisconsin Avenue, Suite 300
Bethesda, MD 20814

ISBN-13: 978-1-4334-0072-8
ISBN-10: 1-4334-0072-3

Printed in the USA

10 9 8 7 6 5 4 3 2 1